AWESOME ACHIEVERS IN TECHNOLOGY

AWESOME ACHIEVERS IN TECHNOLOGY

BY ALAN KATZ

ILLUSTRATIONS BY CHRIS JUDGE

RP|KIDS
PHILADELPHIA

To all of the teachers in my life; they're true Awesome Achievers.

Running Press Kids
Hachette Book Group
1290 Avenue of the Americas, New York, NY 10104
www.runningpress.com/rpkids
@RP_Kids

Printed in the United States of America

First Edition: August 2019

Published by Running Press Kids, an imprint of Perseus Books, LLC, a subsidiary of Hachette Book Group, Inc. The Running Press Kids name and logo is a trademark of the Hachette Book Group.

The Hachette Speakers Bureau provides a wide range of authors for speaking events. To find out more, go to www.hachettespeakersbureau.com or call (866) 376-6591.

The publisher is not responsible for websites (or their content) that are not owned by the publisher.

Print book cover and interior design by Christopher Eads.

Library of Congress Control Number: 2018951803

ISBNs: 978-0-7624-6336-7 (paperback), 978-0-7624-6335-0 (ebook)

LSC-C

10 9 8 7 6 5 4 3 2 1

TABLE OF CONTENTS

A NOTE FROM ALAN KATZ

The way I see it, there are three types of heroes.

There's the fictional, cape-wearing, web-shooting kind. Very exciting, but also very not real.

There's the real-life, military, nation-saving type.

Worth saluting, to be sure.

And then there's a giant list of heroic figures that are unknown to most people. Top achievers whose work hasn't been properly celebrated. They're people who've given us important inventions or technological developments we may have taken for granted. I'm absolutely fascinated by folks who have done so much but have generally lived without fanfare. That's why I'm about to give these people the attention and praise they deserve.

Of course, I should also tell you this: when I find out about one of these Awesome Achievers, my mind forms a ton of questions. Questions such as . . .

How did they do what they did?

What would it be like to be them?

How would I have handled the challenges they faced?

Then, I come up with some answers.

I'll tell you about many of these people on the following pages. And I'll share some thoughts about their accomplishments and how they've changed my life.

Warning: some of my thoughts might get pretty outrageous. But, hopefully, you'll find these people and their very real discoveries as fascinating as I do. And, hopefully, you'll laugh along with me.

Enjoy . . . and thanks!

—ALAN

0
4
PONG

MEET NOLAN BUSHNELL

If there's ever a choice between playing video games and reading a book, I'd take the book every time. But I still admire the creative geniuses behind video games, especially Nolan Bushnell, who's widely considered the Father of Electronic Gaming.

You might say that Mr. Bushnell was a born tinkerer and inventor; he was the kind of kid who would have been perfect for *Shark Tank,* had it existed in the 1950s and 1960s. As a teenager, he developed a roller-skate-mounted liquid fuel rocket in his garage. He also had a successful television repair business at that time, and he later worked at an amusement park while getting an electrical engineering degree at the University of Utah.

Not long after graduation, Mr. Bushnell moved to Silicon Valley, a region in the San Francisco Bay Area that is home to many of the world's biggest technology companies. In 1970, he and a man named Ted Dabney designed and marketed *Computer Space,* the first commercial video game. And a year later, Mr. Bushnell and Mr. Dabney co-founded Atari.

Yes, Atari—the company responsible for classic video games such as *Space Invaders* and *Pac-Man.* Atari started simply, with a 1972 worldwide sensation known as *Pong.* You've probably seen the game—it has a paddle on the left side, a paddle on the right side, and a "ball" moves back and forth on the TV screen. The object is not to let the ball get by your paddle. While it might not seem like much of a game, in the early 1970s, *Pong* was all the rage—first in arcades, and then in at-home versions. It's now on permanent display at the Smithsonian National Museum of American History in Washington, DC.

The *Pong* at-home machine only played that one basic game. Mr. Bushnell and his team followed up with the Atari 2600, a console that accepted interchangeable game cartridges (more than 30 million consoles were sold following its 1977 introduction). Interestingly, it didn't come with *Pong.*

Mr. Bushnell not only developed video consoles and games—including *Centipede* and *Asteroids*—but he also mentored and helped launch the careers of some of the biggest names in the tech industry. Ever hear of Steve Jobs and Steve Wozniak, the men behind Apple? Before starting that company, they both worked for Mr. Bushnell, and they are credited with creating the Atari game, *Breakout*. (Mr. Jobs and Mr. Wozniak later offered to sell a one-third share of Apple to Mr. Bushnell for $50,000. He turned them down, and today, that stake would be worth over $330 billion.)

After selling Atari to Warner Communications, Mr. Bushnell worked in other areas of electronic media. He also founded a *place* to play video games—Chuck E. Cheese's. That's right; Mr. Bushnell didn't invent pizza, but he did dream up and open the restaurant chain where kids can have a slice . . . and a slice of interactive fun!

Without Mr. Bushnell's technological wizardry, it's quite possible that you'd be living life without a PlayStation®, Wii™, or Xbox console. So, the next time you play a video game, say thank you to Mr. Bushnell. (Then put the video game down and read another book!)

PLAYING AROUND . . .

- Over $36 billion was spent on video games in a recent record year. If you spent that much at Chuck E. Cheese's, you could buy 1.44 billion one-topping pizzas, 2.88 billion soft drinks, and 57.6 billion game tokens!

- According to experts, 67 percent of American households own a video game device. And video gamers ages thirteen and older play more than six hours a week. If you *read* more than six hours a week, your brain will get smart and your thumbs won't ache (except maybe from turning the pages).

- Beginner players burn up to 350 calories during an hour of real table tennis (the kind with a real bouncing ball that your sister usually steps on when she's losing). An hour of playing *Pong* burns about 80 calories.

- Some Atari game cartridges are worth a ton of money. Several years ago, a new, in-box copy of *Air Raid* sold for over $33,000!

WE HAD A BALL

When I was a kid, I used to play *Pong* with my dad every night at 7:30. We'd sit together on the couch (the controller knobs were *built onto* the console, so you had to sit right next to each other to play), and he'd win game after game after game. If we played to 11, he'd win 11–3 or 11–4. If we played to 21, he'd win 21–3 or 21–4. But then something very strange would happen . . .

At about 7:59, I'd manage to win a game—usually by the lopsided score of 11–0 or 21–0.

I'd brag about how I'd worn him down, and how he was finished (finished!) as a *Pong* player. I'd go to my room happy and tackle my homework assignments with a smile on my face.

Only now do I realize that he was *letting me* win that last game, so he could watch his favorite 8 p.m. TV show and get me to do my homework without putting up a "let's play some more" fuss.

How did I not catch on to his trick night after night, month after month? I really don't know. And you might think my favorite time of all was 7:59, when I got to win a game. But it wasn't; it was 7:30, 'cause that's when we always sat down to play.

TV . . . OR NOT TV?

Nolan Bushnell is credited with giving the world video games. But his devices weren't the first time a television screen was used for interactivity. In the mid-1950s, there was a TV show called *Winky Dink and You* that gave a kid the chance to participate with on-screen animated characters by buying a plastic, erasable "magic drawing screen" and affixing it to his or her TV set. Using special crayons, that viewer could draw a bridge to help characters escape danger, solve mysteries, and get involved in other ways. It was quite popular back then, though parents did complain for two reasons: one, because they were afraid that sitting so close to the TV would harm their kids, and two, because kids who didn't own the plastic screen kit would just draw right on their television screens.

The show lasted from 1953 to 1957, and during that time, kids everywhere learned something that you surely know by now: never draw on a TV screen!

MEET ADAM CHEYER AND DAG KITTLAUS

If you—or family members—have an iPhone, an iPad, Apple Watch, or a Mac computer, chances are you're familiar with the artificial intelligence wonder known as Siri. Siri is a conversational personal assistant, always ready to supply information about the weather, sports scores, directions, jokes, and so much more.

Having Siri with you is better than having a know-it-all friend by your side. Quite simply, "she's" amazing.

As with any good invention, Siri started with a bold vision and a desire to fill a need—in this case, a need for a system that would understand spoken language and assist with scheduling, maps, reminders, and more.

Adam Cheyer, a Rubik's Cube champion as a young boy (his average time for solving the cube was twenty-six seconds!), was an enthusiastic computer programmer who joined SRI, a California research institute that had given the world the computer mouse and inkjet printing, among other innovations. (Although it looks like the word *Siri*, SRI, in fact, stands for Stanford Research Institute.) In the early 2000s, Mr. Cheyer went after his goal, which he later said was to build "a humanlike system that could sense the world, understand it, reason about it, plan, communicate, and act."

Mr. Cheyer and his team at SRI developed what they called a Cognitive Assistant that Learns and Organizes, or CALO for short. About five years later, Dag Kittlaus joined SRI as an entrepreneur-in-residence. Mr. Kittlaus had grown up in Illinois, but he got his degree at BI Norwegian Business School in Oslo, Norway. (His mother was a native of Norway and Dag frequently visited the country while growing up.) After starting his technology career in Norway, he returned to the United States and spent five years working at Motorola, a telecommunications company.

In his early days at SRI, Mr. Kittlaus suggested that CALO would be a great addition to the just-released iPhone. The team needed to devise a system that would recognize human speech and employ its own local search engines—to help it understand whether you wanted to drive to, say, Springfield, Illinois, or Springfield, Massachusetts. They also gave Siri the capacity to handle personal data, accept credit card information, and more, all with a dependable level of accuracy. After about two years of development, funded by $25 million in investments (and aided by artificial intelligence expert Tom Gruber and engineer Chris Brigham), the system was ready to go; it was renamed Siri and it was introduced in Apple's App Store in February 2010.

The team thought they had produced something great. And that thought was confirmed just a few weeks later, when Steve Jobs, the chairman and co-founder of Apple, called Mr. Kittlaus and invited him to visit him the next day for a meeting. Mr. Kittlaus responded, "No, I can't. I'm taking my wife to the ballet." Mr. Jobs said, "Really? Then you could come over before."

And so, he did. They met, and Mr. Cheyer and Mr. Kittlaus turned down Mr. Jobs's offer to buy Siri. But Mr. Jobs was persistent, and before long, Siri, the world's first virtual assistant, was sold to Apple for an estimated $200 million. Was Siri worth that much money? Don't ask me—ask *her*.

NO, SIRI DIDN'T NAME HERSELF...

It's said that Siri was so named because Mr. Kittlaus planned to name his daughter Siri. (In Norwegian, it's a popular girl's name that means "beautiful woman who leads you to victory.") But he and his wife had a boy, so he decided to use the name for his new technology. Some people consider the letters in the word *Siri* to represent "Speech Interpretation and Recognition Interface."

Siri is a fine name. But I would have named her Aunt Selma. Why? Because I have an Aunt Selma and she knows *everything*. No matter what you ask her, she knows the answer. She's incredible. She's received trophies as the winner of a spelling bee, a geography bee, and a math bee, and it's like she has a GPS built into her brain. She always knows exactly how to get somewhere, even if she's never been there before. Her storm prediction abilities outdo those of any TV meteorologist. She diagnoses ailments better than any doctor, and she's never even taken a medical course. She can name any dog breed, can compute any series of complex numbers without using a calculator, and she speaks twelve languages. You should meet her!

HELLO
CIAO
HOLA
BONJOUR
SZIA
HABA
ALOHA
OHAYO
NIHAO

SHE SAID IT ALL!

Susan Bennett is the original voice of Siri (since 2013, iPhone owners have had the option to switch to a male voice). Ms. Bennett was hired to record phrase after phrase for weeks on end, never knowing what purpose her recordings would serve. When Apple acquired Siri and introduced the system on their iPhones, Ms. Bennett was completely surprised.

I wonder: When Susan Barrett uses her iPhone, does she say, "Hey, Siri?" Or . . .

"Hey, *me*?"

I'M GLAD I ASKED

Here are some *actual* responses to things I said to Siri:

"Do you have any dessert?"

"I don't eat. But if I did, I'd eat chocolate."

"Can I borrow some money?"

"Alan, you know that everything I have is yours."

"Will you do my homework?"

"The answer is the square root of 1,764."

"I think you are very smart."

"You can't tell, but I'm blushing."

"Do you sleep?"

"I never rest. But thanks for asking."

I LOVE HIM TO THE SPOON AND BACK

Way before Siri was introduced, I had a personal assistant.

I didn't have to own a phone or press a button. And it wasn't a *virtual* assistant.

It was my three-year-old son Andrew, who was a willing (if not always correct) helper. I'd say . . .

"Andrew, bring me a spoon."

He'd run away, and come back with a pointy, shiny object.

"Here, Daddy."

"Thank you, Andrew, but remember, the spoon has an oval shape at the end."

"Oh, yeah."

He'd run away, and come back with a spoon.

"Thank you, Andrew. Here's a quarter." *(Actually, it was a penny.)*

"Cool, Daddy. Thanks!"

My assistant is now a college graduate. And I've learned an important lesson: if you write about all the times you gave a little kid a penny and told him it's a quarter, when he grows up, he's going to read about it—and ask for the other twenty-four cents. I did that about one thousand times . . . so, I owe him $240 dollars.

I should have gotten my own spoons.

MEET NILS BOHLIN

You use it dozens of times a week. And without question, it saves your life each time. It's the automobile seat belt. More specifically, the three-point lap and shoulder seat belt.

Until 1958, cars only had two-point lap belts that just went around the waist. But they failed to protect the chest area of the driver and passengers and often caused severe body and head injuries when a crash occurred.

Enter Nils Bohlin, the Volvo Car Corporation's first chief safety engineer.

Born in Härnösand, Sweden, Mr. Bohlin earned a degree in mechanical engineering. He then worked for

Saab as an aircraft designer, developing ejector seats and pilot rescue systems for the company's fighter airplanes.

In 1958, Mr. Bohlin joined Volvo, where his previous experience served him well in designing the three-point seat belt (although airplanes used four-point belts, he knew that such a design wouldn't be appropriate for automobiles). Mr. Bohlin found a way to secure the upper and lower body, with the straps holding them safely by buckling into what Mr. Bohlin referred to as "an immovable anchorage point" below the hip. According to Mr. Bohlin, "It was just a matter of finding a solution that was simple, effective, and could be put on conveniently with one hand."

Mr. Bohlin's insistence on one-hand usage was critical to his design. He said, "The pilots I worked with were willing to put on almost anything to keep them safe in case of a crash, but regular people in cars don't want to be uncomfortable even for a minute."

In 1959, Volvo introduced the three-point seat belt in its automobiles. In an effort to maximize automobile safety, Volvo offered the innovative design to other car manufacturers—at no cost. (That's pretty rare; generally, companies competing for sales and customer loyalty will keep such important breakthroughs to themselves.) It's estimated that injuries sustained from crashes declined by 90 percent with the introduction of Mr. Bohlin's invention, and that well over one million lives have been saved since the three-point seat belt was first installed in cars.

Mr. Bohlin worked at Volvo until 1985, and he was inducted into the Automobile Hall of Fame in 1999. He was also welcomed into the National Inventors Hall of Fame in Akron, Ohio, on the day he died in 2002.

"We believe that he was a great inventor; an inventor with a conscience that made great contributions to road safety," said Victor Doolan, chief executive and president of Volvo Cars of North America. "There is a little bit of Nils Bohlin in every car."

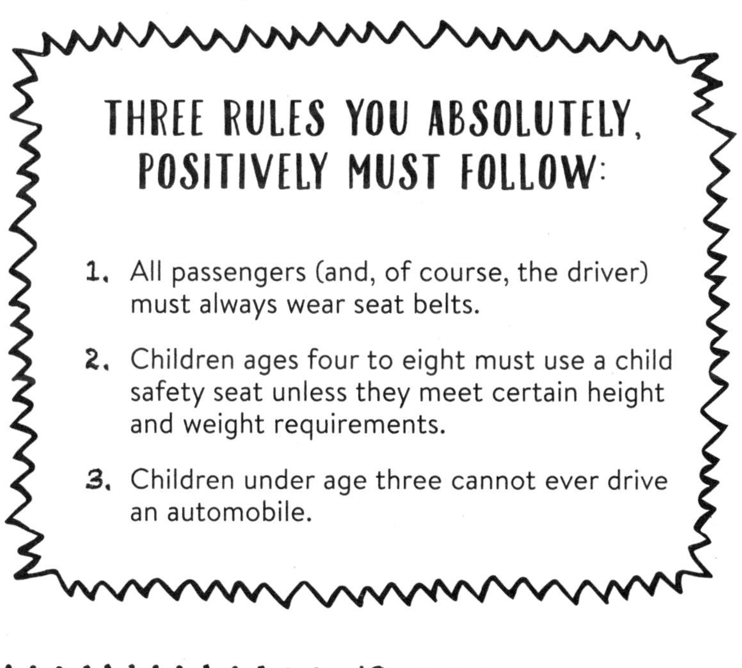

A POEM I WROTE WHEN I WAS SEVEN YEARS OLD:

When my brother got out of our daddy's car,
he put the family in a trance.
See, when he unbuckled his safety belt,
he'd also unbuckled his pants!

I think seat belts are so important, I spent some time coming up with slogans to tell everyone about Mr. Bohlin's invention. What do you think of these?

IF YOU BUCKLE UP FOR SAFETY, I WILL YELL, "HIP-HIP-HOORAFETY!"

BUCKLE UP, 'CAUSE THAT'S THE LAW . . . AND I'LL GIVE YOU SOME TASTY COLE SLAW!

IF YOU FORGET TO BUCKLE, YOU'LL SURELY BE IN TRUCKLE.

REMEMBER TO BUCKLE YOUR SEAT BELT . . . I ASKED THE BUTCHER HOW RAW MEAT FELT.

GRAB THE BUCKLE, MAKE IT SNAP, OR ELSE YOU ARE JUST FULL OF CR–

Never mind. It's hard to write a slogan. But it's easy to remember this:

BUCKLE UP EACH AND EVERY TIME YOU'RE IN A MOVING VEHICLE!

Hey, *that's* not a bad slogan!

MEET ROBERTA WILLIAMS

Not every Awesome Achiever set out to change the world. In fact, video game creator Roberta Williams, often called The Queen of Graphic Adventure," had no thoughts of being such an innovator.

Ms. Williams was a shy child with a wild imagination. She often enthralled her family members by creating fairy-tale adventure stories. But she probably never could have imagined what the future held for her.

In 1979, Ms. Williams was a mom of two and knew very little about video games and computers. That year, her husband Ken (a computer programmer and consultant) showed her a simple game called

Colossal Cave Adventure, in which players had to type text directions in order to control characters. Ms. Williams loved *Colossal Cave Adventure* so much, she bought and played additional text-based adventure games. And before long, confident that she could do an even better job, she set out to create a game of her own.

But Ms. Williams was a creative storyteller, not a computer expert. And so, she persuaded her husband to develop a program based on her tale, which she called *Mystery House.* The game was revolutionary, and here's why: it enhanced the players' enjoyment by including simple black-and-white graphics (something that *Colossal Cave Adventure* and other early games didn't have).

After a month of programming, the game was ready; it was released in May 1980. After a mere three months, the couple had earned $60,000 on *Mystery House.* Their follow-up game that year, *Wizard and the Princess,* featured color graphics and was also tremendously popular.

Mrs. and Mr. Williams started their own company, On-Line Systems, in late 1980. The company later became known as Sierra Online, and they released more and more successful games, including *Mission: Asteroid, Time Zone,* and *The Dark Crystal* (based on a 1982 adventure-fantasy movie produced by the Jim Henson Company).

Soon after, Ms. Williams began creating what would be her most beloved project: *King's Quest*. That game was the first to feature animation-like characters that the player could control, moving them around in an environment without typing. An incredible blockbuster, *King's Quest* became one of the most popular adventure games ever released, and it led to seven sequels, as well as three motion pictures.

Other successful titles followed, including *The Black Cauldron*, *Mickey's Space Adventure*, and *Mixed-Up Mother Goose*.

Ms. Williams has been heralded as one of the most respected (and successful!) computer game creators, and it's notable that she is widely considered the first to make many of her games' main characters female.

"I am most proud of the development of the characters as personalities that game players could relate to and care about," Ms. Williams once said.

Game players enjoyed her games. Many still do. And it's all thanks to a little girl with a wild imagination who grew up to share it with the world.

I TRIED, I TRIED...I REALLY TRIED

Ms. Williams was so inspired by playing games that she wanted to create some of her own. That's absolutely incredible. I like to think I'm creative, but I certainly don't have that kind of inventiveness. Still, based on Ms. Williams's achievements, I came up with my own game ideas. They all failed.

Sonic the Groundhog Watch him pop up! Fun for the whole family. But the game could only be played on February 2.

Snail Kong Like *Donkey Kong*, but with a snail as the main character. Unfortunately, snails move at a snail's pace (slowly!), and the game took 419 hours to play.

Spice Invaders Someone's raided the spice rack! The goal: to find the missing salt, pepper, oregano, and garlic! The problem: no one wants to chase after garlic.

Grand Theft Pencil Jimmy's pencil has been stolen, and the player had to get it back before the big math test. Unfortunately, the game—like the pencil—had no point.

Call of Doody I don't think I need to explain this one; let's just say there was a lot of explosive action.

But I don't give up easily; after my lousy video game titles, I turned my attention to other inventions that I thought were genius. But they were lousy, too (maybe even lousier than my game ideas):

Drink coasters made of salami and bologna slices. You could protect the table by using the slice to put your beverage on, then eat the salami or bologna. It was useful . . . and delicious. (This idea didn't work because after you ate the slice, you had to put your beverage on the table anyway.)

A combination barber shop and ice cream store that was meant for little kids who tend to squirm while getting a haircut. When the toddler sat down, he or she got a huge ice cream cone to enjoy as the barber did his or her snipping. The kid was distracted and sat quietly. Perfect, right? Wrong. Turns out no kid wants to eat an ice cream cone with their just-cut hair all over it, no matter how hard you try to convince them that the hair is just thin sprinkles.

Movie theaters with popcorn poppers right at the seats. No need to get up—your fresh snack is right where you're sitting. A great idea, but when the popcorn popped, you couldn't hear the movie. And don't even ask about the butter . . .

Pop-up cheese slices. Rather than unwrapping them individually, I came up with a way to have cheese slices pop up, much as tissues in a box do. Worked great, until people thought they *were* tissues and started blowing their noses in the cheese slices.

IMPORTANT MATH! IMPORTANT MATH! IMPORTANT MATH!

If you play video games 1 hour a day, every day, between the ages of 8 to 18, that's about 365 hours a year for 10 years (including leap years), which equals 3,650 hours. A total of more than 150 days of your life. So . . . DON'T!

THE ULTIMATE GAME CONTROLLER!

(Sorry, size XX3CCm17rpq batteries not included.)

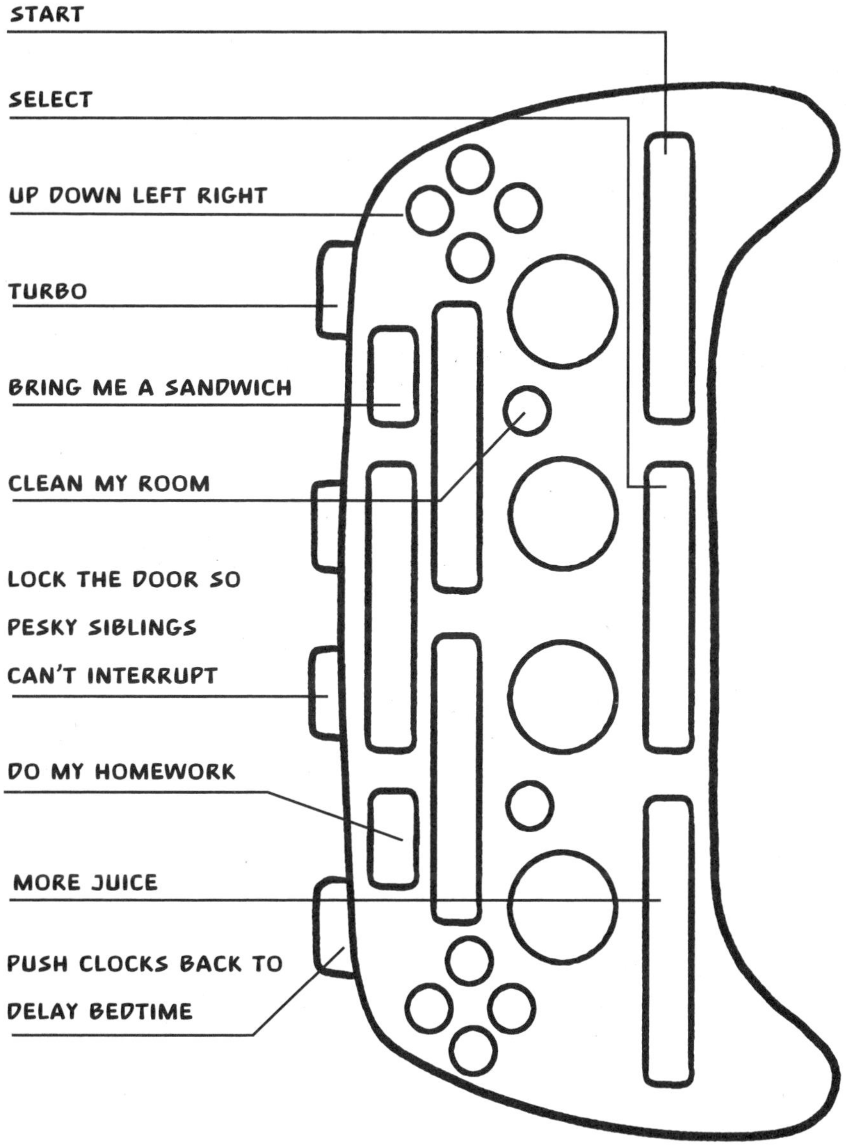

START

SELECT

UP DOWN LEFT RIGHT

TURBO

BRING ME A SANDWICH

CLEAN MY ROOM

LOCK THE DOOR SO
PESKY SIBLINGS
CAN'T INTERRUPT

DO MY HOMEWORK

MORE JUICE

PUSH CLOCKS BACK TO
DELAY BEDTIME

MEET ROBERT ADLER

Ever hear the term *couch potato*? That's what someone's called when they simply lie on the couch watching TV, perhaps for hours on end.

Well, Robert Adler didn't develop that phrase; rather, that phrase was created because of something he developed the perfectly functioning wireless TV remote control.

In the very early days of television (the late 1940s), there simply was no way to control the TV from the couch. If you wanted to turn it on or off, or change the channel, or adjust the volume, you'd have to get up and do it by hand.

Then in 1950, a company known as Zenith introduced a device called Lazy Bones. This device actually featured a wire that was attached to the TV—to literally connect the viewer to the television. A motor in the TV operated the channel changer dial via the remote control. Imagine how exciting that was to (lazy) viewers, as the product's advertising boasted, "Prest-o! Change-o! Just Press a Button . . . to Change a Station!"

A Zenith engineer later invented the Flash-Matic, which was wireless and used a precise flashlight to control the main functions of the TV. But there were problems with the device: if sunlight hit the dials on the TV, it would change channels all by itself. So long, Flash-Matic.

In 1955, while working at Zenith, Mr. Adler then stepped in and decided that the answer to remote control technology was to employ ultrasonic tones—high-frequency sounds that were beyond the range of human hearing. A year later, the Zenith Space Commander was introduced. The device was a true wireless remote control. And guess what—because it used aluminum rods that were struck by small hammers when keys were depressed, the remote control didn't need batteries!

By the way, because the Zenith Space Commander created the need for an additional receiver inside the TV, the price of a set went up by about 30 percent. But consumers didn't care—they *wanted* to be couch potatoes!

The technology behind Dr. Adler's remote control was the standard in the television industry until 1980; in its twenty-five years of use, more than nine million such units were sold. After that, remote controls were produced with infrared (or IR) technology. Rather than making sounds the human ear couldn't hear, IR remotes generate lights the human eye can't see. IR technology is used to this day.

Mr. Adler worked with Zenith for more than six decades; over that time, he earned more than 180 U.S. patents, including one for the gated-beam tube, which improved TV sound reception.

You're probably wondering: was Mr. Adler himself a couch potato? Not at all. According to his wife, he didn't watch a great deal of television. "He was more of a reader," she said. "He was a man who would dream in the night and wake up and say, 'I just solved a problem.' He was always thinking science."

WOW, THIS IS VITAL TO KNOW!

If you rearrange the letters in the words *REMOTE CONTROL*, you can spell *ELECT NORM ROOT*. Who's Norm Root? And in which election is he a candidate? I have no idea. But I thought you'd find it interesting. You don't? Oh, sorry.

GET THE COUCH POTATO BACK WHERE HE BELONGS!

MORE IMPORTANT MATH! MORE IMPORTANT MATH! MORE IMPORTANT MATH!

If the remote control had never been invented, Americans would be in much better physical condition. Consider this: If the average couch is 6 steps away from the TV, that means the viewer would have to walk a total of 12 steps to change the channel. And since there are approximately 2,000 steps in a mile, that means that getting up to change the channel 167 times would cause the viewer to walk a mile. And walking a mile burns about 100 calories—so based on how often they change the channel in our house, my kids would lose about 42 pounds a day going roundtrip from couch to TV. Thanks a lot, Mr. Adler!

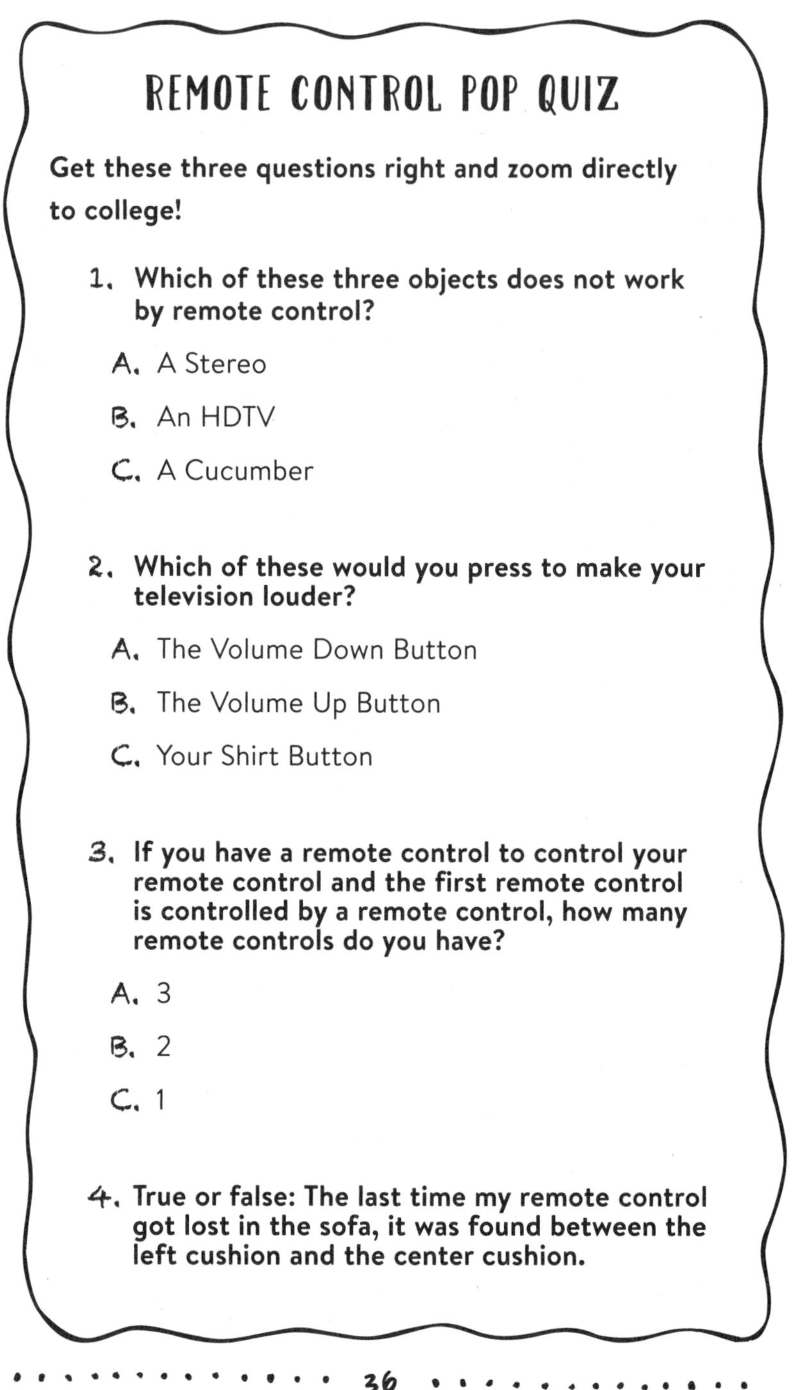

REMOTE CONTROL POP QUIZ

Get these three questions right and zoom directly to college!

1. Which of these three objects does not work by remote control?

   A. A Stereo

   B. An HDTV

   C. A Cucumber

2. Which of these would you press to make your television louder?

   A. The Volume Down Button

   B. The Volume Up Button

   C. Your Shirt Button

3. If you have a remote control to control your remote control and the first remote control is controlled by a remote control, how many remote controls do you have?

   A. 3

   B. 2

   C. 1

4. True or false: The last time my remote control got lost in the sofa, it was found between the left cushion and the center cushion.

Answers: 1 (C), 2 (C), 3 (A), 4 True

MEET MARY ANDERSON

How would you feel if I told you that one of the most technological advancements in our lives was something that was invented back in the early 1900s?

It's not computerized. It doesn't make or receive calls, and it won't help you even a little at homework time.

It's . . .

. . . the windshield wiper.

That's right; back in 1903, Mary Anderson received a U.S. patent for what she called "a window cleaning device for electric cars and other vehicles to remove snow, ice, or sleet from the window." (The patent was issued to protect her idea from being stolen by others.)

It all started on a snowy, icy day in New York City. Ms. Anderson was a passenger on a city trolley car, and she noticed that the driver had seeing through the window. In fact, he had to keep sticking his head out of the side window—or get out of the trolley car and clean the snow from the windshield—to see where he was going.

It's said that Ms. Anderson began to sketch an idea right there on the trolley car, conceiving what we now call windshield wipers. Her initial design featured wood and rubber wiper arms attached to a lever near the streetcar's steering wheel. Pulling the lever caused the spring-loaded arm to drag back and forth across the window, thus clearing any precipitation. And she made the wipers removable, because she reasoned they weren't needed when the bad weather season ended.

Sadly, people laughed at Ms. Anderson's invention; they thought the movement of the wipers would be distracting and would possibly even cause accidents. Besides, since early vehicles didn't travel at today's high speeds, many didn't even have windshields! Ms. Anderson's patent expired in 1920, and as the number of personal cars on the road increased, many companies later copied her idea (once the patent had expired, others were free to create similar products).

Ms. Anderson never succeeded in marketing her

innovation, which is ironic considering today's cars cannot legally be driven without them.

Reverend Sara-Scott Wingo, Ms. Anderson's great-great niece, had this to say: "We're all really proud of her. I have three daughters. We talk about Mary Anderson a lot. And we all sort of feel like we want to be open and receptive to sort of our own Mary Anderson moments."

Mary Anderson moments. Brilliance out of the blue. What a nice thought. Put your thinking cap on and try to have at least one today.

WHO NOSE WHERE GREAT IDEAS COME FROM?

After researching Ms. Anderson's story, I felt very inspired. And that very week, I was riding on a New York City bus (they don't have trolley cars anymore). It was a rainy day, and the bus driver was using her windshield wipers; they were greatly improving her visibility. I nodded to the man sitting across from me, as if to acknowledge that Ms. Anderson's invention was doing a terrific job.

He had apparently never heard of Ms. Anderson, and he didn't nod back. In fact, he sneezed. And I immediately began to sketch an idea right there on the bus. After getting off the bus two miles past my stop (that's how excited I was!), I applied for a U.S. patent for what I called "a nose cleaning device for sloppy people to remove mucus from the face, shirt, and pants."

The woman at the patent office said my invention was nothing to sneeze at. I took offense and told her yes, it was. She said, "No, it's not. 'Nothing to sneeze at' means I think it's a good idea." I said thank you, but she still refused to give me the patent. Here's my design:

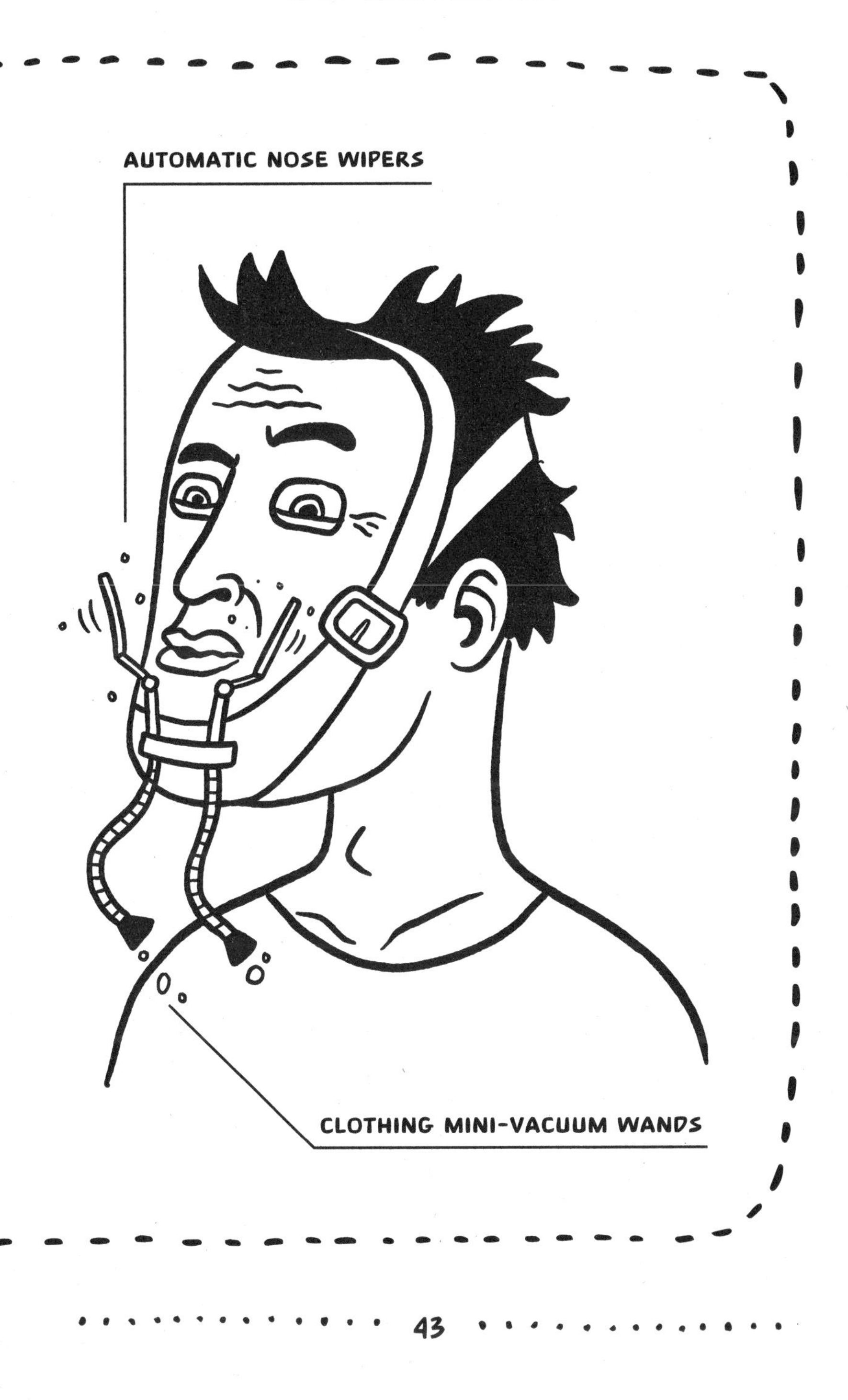
AUTOMATIC NOSE WIPERS
CLOTHING MINI-VACUUM WANDS

I'm going back to the U.S. patent office with an even ~~smarterer~~ ~~brillianter~~ better idea that will put windshield wiper companies out of business! It's my . . .

Car Umbrella!

To test it, I got the biggest umbrella I could find. My wife drove the car, and I opened the sunroof and stuck the giant umbrella up through it. I opened the umbrella and the car stayed dry for the whole trip! Not a drop of water on the car!

Amazing, right?

Well, it's not *that* amazing. See, it actually wasn't raining that day.

ON (FEBRUARY) 2ND THOUGHT ...

I've always known that whenever we set the clocks ahead or back for daylight saving time, it's also a good idea to change the batteries in our home smoke alarms. That's certainly a good way to stay on top of an important task.

But it was only while researching the life of Mary Anderson that I learned that in the United States, February 2 is National Change Your Windshield Wipers Day. Why February 2? It's Groundhog Day, and it's said that whether the groundhog sees his shadow or not, new wipers will properly prepare your car for winter weather or spring showers. (Of course, it's not necessary to wait until February to change your wipers if they get streaky or smeary at other times of the year.)

MEET MARTIN COOPER

When I was a kid, telephones were wired to the wall; you could walk and talk as far as that wire would allow. And outside of the house, if you needed to make a call, you had to use a pay phone. That meant you needed to have coins to drop into the slot, and often, you had to have patience as well—because the person using that pay phone was gabbing and chatting and blathering and jabbering . . . on and on and on.

Personal cell phones? They were only mentioned in sci-fi conversations.

Martin Cooper changed all that.

An electrical engineer who served in the U.S. Navy

during the Korean War, Mr. Cooper joined the Motorola Corporation in 1954. There he worked to introduce the first radio-controlled traffic-light system, as well as the first handheld police radios. He also built the first electronic device for enabling a telephone to ring in the car.

In the early 1970s, Mr. Cooper led Motorola's team in the development of the DynaTAC phone, a portable cell phone that weighed two-and-a-half pounds and was nine inches tall (the name stood for Dynamic Adaptive Total Area Coverage). The phone was heavy (and gigantic) by today's standards, and its battery only allowed thirty-five minutes of talk time on a charge.

On the day that the DynaTAC was introduced at a New York City press conference, Mr. Cooper made the first public cell phone call ever—to brag about the new phone to Joel Engel, head of a similar project at AT&T, Motorola's chief rival. His immortal words: "I'm calling you from a cell phone . . . a handheld, personal portable cell phone."

The phone stayed in development for another ten years; in 1983, Motorola introduced the more advanced DynaTAC8000X, the first consumer portable cell phone (its price was about $4,000!).

Since then, Mr. Cooper's breakthrough technology

has been adapted and improved by many companies. Today's mobile phones are also cameras, video recorders and game machines, with fitness and health technology, calendars, and so much more built in. And, best of all, they're much smaller and weigh considerably less than the DynaTAC.

As for Mr. Cooper, he went on to play an important role in the introduction of Wi-Fi, liquid-crystal displays, and two-way pagers. He is widely considered a visionary leader in wireless technology. Mr. Cooper was quoted as saying, "I'm not that great an engineer, but I'm a very good dreamer . . . the skill is ignoring reality and thinking about what could really be."

Mr. Cooper, thank you for dreaming big. With more than half of the people in the world owning cell phones and using Wi-Fi daily, your dreams have changed lives everywhere.

Wi-Fi is not an abbreviation; it's a made-up term that doesn't actually stand for anything. And by the way, the same is true for *Häagen-Dazs* (the words don't translate into any language).

IT'S NOT WHAT YOU SAY, IT'S HOW FAST YOU SAY IT

The first mobile phone I ever bought weighed a couple of pounds. But it wasn't the heaviness of the phone that upset me, it was the fact that I paid about fifty cents for every sixty seconds I used it (there were no unlimited calling plans then, and you paid that much whether you made or received the call). And even if you spoke for one second over the first sixty, both the caller and the recipient would each be charged another fifty cents.

When I'd call my wife's mobile phone from my mobile phone, we'd always have a speedy conversation. Like this . . .

"Hihowareyou?"

"I'mfinehowareyou?"

"I'mfinetoo. Whatdoyouwantfordinner?"

"Ziti." *(It took too long to say "spaghetti.")*

"Metoobye. I'llseeyoulater."

"Okaygreat. Bye."

Yes, we'd pay a total of one dollar for that conversation. Which, when you think about it, is probably more than the ziti cost us.

CONNECTION CORRECTION

Cell phones can be important when you need a ride or when you need to find out on which field the soccer game is being played. But they can also be annoying . . . which was the case with the call I recently got from Jessie:

Jessie: Hi, this is Jessie. Can I talk to Jordan?

Me: *May* I talk to Jordan?

Jessie: I'm sorry. *May* I talk to Jordan? I heard he wasn't feeling good.

Me: You heard he wasn't feeling *well*.

Jessie: I'm sorry. May I talk to Jordan? I heard he wasn't feeling *well*, and Sue and me want to see how he's doing.

Me: Sue and *I*.

Jessie: I'm sorry. May I talk to Jordan? I heard he wasn't feeling well, and Sue and *I* want to see how he's doing. I hope he's better then yesterday.

Me: Better *than* yesterday.

Jessie: I'm sorry. May I talk to Jordan? I heard he wasn't feeling well, and Sue and I want to see how he's doing. I hope he's better *than* yesterday.

Me: Very well said. But . . . there's no Jordan here. You've got the wrong number. Good-bye.

MEET MARIE VAN BRITTAN BROWN

It was 1966. Nurse Marie Van Brittan Brown lived in Jamaica, New York, with her husband, Albert, who was an electronics technician. Because Mr. Brown worked during the day and Ms. Brown worked at night, she worried about her personal safety, and she always wanted to know who was at her door when she was home alone.

Not satisfied with the traditional front door peephole, Ms. Brown—who was not an inventor at that point—conceived the first home security system. Albert's knowledge of electronics came into play and here's what they devised:

Multiple peepholes were drilled into the front door. A

cabinet was affixed to the inside of the door, and in the cabinet was a camera that could be moved up and down remotely to scan the peepholes and send a video image of the visitor to a monitor in the bedroom (the camera moved up to get a look at tall visitors, and down to look at shorter people).

So, in the comfort and security of her bedroom, Ms. Brown could see who was at her front door. The bedroom technology also included a two-way microphone and speaker so that Ms. Brown could communicate with the person at the door, and featured buttons so that she could remotely lock the door or sound an alarm.

Three years later, Ms. Brown and her husband received a patent for her ingenious device, then called a "Home Security System Utilizing Television Surveillance."

What Ms. Brown had invented was a means of feeling secure at home. But in doing so, she actually pioneered the use of closed-circuit television—with an audiovisual signal sent to a specific monitor (as opposed to broadcast TV, which offers pictures and images that millions could receive and watch at once).

This provided the foundation for security camera systems now used to monitor visitors to countless airports, banks, and shops. And of course, though more modern in shape, form, and functionality, they're still widely used in homes and apartments around the globe. Today, with iPhone and iPad technology plus an additional device, you can see who's at your front door and communicate with that person even if you're *not* at home. And though you may not have to stand behind a closed door and ask, "Who's there?," now you know who was there *first* . . . Marie Van Brittan Brown.

THE JOKE'S ON YOU

Though I'm sure Ms. Brown had a good sense of humor, I'd also bet she didn't enjoy hearing knock-knock jokes. I'd imagine if you'd tried to tell her one, it'd go something like this . . .

You: Knock, knock.

Ms. Brown: I already *know* who's there. I can see you on the monitor in my bedroom. Please go tell that joke to someone else!

GO AWAY!

A CLOSED CIRCUIT SENSATION

When my daughter was a little girl, I'd take her grocery shopping every Saturday morning. As we entered the market, she'd see herself on the monitor and we'd have to stop for a few minutes so that she could "do a show."

I'd say, "Live, from Peter's Market . . . it's the Simone Show!"

She'd sing. She'd wiggle out of the shopping cart and dance. She'd do magic tricks, such as making our shopping list disappear by crumpling it up and throwing it off camera.

People would walk by and tell me how cute she was (even if she *was* blocking the entrance). I'd tell them thank you, but she'd let them know that they couldn't see her in person—they had to watch her "show" on the TV monitor.

Her weekly performances were adorable, and frankly, they were much better than most of the shows currently on TV.

Stay tuned. Simone will be right back. And be sure to catch the special sale on onions in the produce aisle!

GREATNESS CONTINUES . . .

Mr. and Ms. Brown had two daughters, one of whom (Norma) became a nurse and an inventor. She also specialized in the area of personal well-being.

I'm sure that in the case of her daughter Norma, Ms. Brown was very pleased that her daughter followed in her footsteps as a nurse and inventor.

I'm very hopeful that someday, one of my kids will follow in my footsteps and replace me . . . as the best author on earth.

IMPORTANT TIP

If you have a security camera, and you notice there's someone at your front door who you don't want to see, just remain silent. Because if you yell, "There's nobody home!" that might give them the idea that there *is* somebody home.

WWW

MEET SIR TIM BERNERS-LEE

On July 16, 2004, London-born Tim Berners-Lee was dubbed a Knight Commander, Order of the British Empire (KBE) by Her Majesty Queen Elizabeth II. He received this honor as the inventor of the World Wide Web.

Sir Berners-Lee wasn't always a "Sir," and he wasn't always an inventor. Wait, I take that back; in a way, he *was* always an inventor. As a child, he had a model railroad set up in his bedroom, and he invented electronic devices to control the trains. Before long, he cared more about electronics than he did about trains. Later, as a college student, Berners-Lee was able to turn a television set into a working computer. (I'm sure he benefitted from having parents who were early computer scientists!)

Berners-Lee's first job was at CERN, a particle physics laboratory outside Geneva, Switzerland. When scientists from around the globe came to visit, he noticed that it was difficult for them to share information. As he put it, "In those days, there was different information on different computers, but you had to log on to different computers to get at it. Also, sometimes you had to learn a different program on each computer."

Yes, the Internet was emerging as a way to connect millions of computers. But between 1989–1990, Berners-Lee created the first web browser. He also wrote the three fundamental technologies that remain the foundation of today's Web; HTML, URL, and HTTP.

As news of the World Wide Web grew, Berners-Lee realized that it would only really take off if it were available to all for no charge, and with no permission necessary for usage. To give away such a valuable invention was a bold, generous step. Indeed, it's been accessible worldwide for free since it was introduced back in 1990.

For his life-changing breakthrough, Sir Berners-Lee was named one of *Time* magazine's "100 Most Important People of the 20th Century." He is also a member of the

Internet Hall of Fame, and he's received more than ten honorary degrees. And if you don't believe me, you can check it out on the World Wide Web.

Parents always say, "When I was a kid, we didn't have the Web. We looked things up at the library." And that's true; we didn't have computers *or* the Web for research. In so many ways, information is much more accessible than it was decades ago. But please remember this: although today we visit the Web numerous times a day for all sorts of facts and fun, there's still plenty of both—and more—at your local library.

IT'S WIDER THAN YOU THOUGHT!

It's said that there are more than 1.3 billion different websites on the World Wide Web.

If you wanted to spend 1 minute on each of those sites, it would take you approximately 2,471 years. If you're gonna try, I'd recommend waking up really early.

I TRIED, I TRIED... I REALLY TRIED (PART 2)

Here are some more (failed) ideas that I was sure would make me a famous inventor:

WWW.com. Yes, I was on the same invention path as Sir Berners-Lee—though *my* idea stood for World Wide Washcloth. It was a lousy, disgusting notion . . . and frankly, there was no way to make it work.

Reusable butter. Spread it on toast, eat it, and then use it again. A brilliant idea, but I kept running into the same problem—after the butter was eaten, it was gone.

Reusable toast. Same exact problem as above.

Clothing made out of grapes. A simple idea: wear it to the beach, and after some time in the sun, you'll have clothing made out of delicious raisins. The idea would have worked, because grapes *do* dry in the sun and turn into raisins. *BUT* . . . it takes weeks, and that's a long time to spend at the beach. And here's an even bigger concern: even if you *did* wait for the grapes to turn to raisins, when you ate them all, you'd be naked. Grapes gone. Raisins gone. Clothes gone. (In many ways, it's like the butter and toast problem.)

Soap baseballs and softballs. Your chance to make a great catch . . . and a great chance to get clean while playing. Bonus: foul balls give fans with sticky cotton candy the opportunity to catch them and get un-sticky right there in the stands. Like Sir Berners-Lee, I offered this idea for free to pro teams and local leagues, and they all turned me down. Sigh . . . it's hard to be a deep thinker.

A book called *Awesome Achievers in Technology*. The problem with that idea was . . . oh wait, there's no problem with that idea. It's a real book, you're reading it, and you're about to jump up and down and say, "This is my favorite book ever!" (Unless you're somewhere that jumping up and down is a bad idea, such as in a car or on a roller coaster.)

IT'S TRUE THAT NOT EVERYTHING IS TRUE

The World Wide Web can be an amazing place, full of valuable news and information. BUT . . . please remember to use it with caution, because not everything you read online is the truth. Just because it's there for you to see doesn't make it factual.

Here's a true example: I visit a lot of schools across America to share my books and my love of reading and writing. I've been to more than 250 different schools in about 30 states. And on one of those trips, I met a North Carolina elementary school teacher who wanted her students to be smart when doing research. So she made a *fake* website full of outrageous historical "facts"—things that were very far from true. Things like, "Columbus discovered America by driving a truck across the country in 1955," and "Texas and New York fought each other in the War of 2003."

She gave her students a list of websites to visit for help with a history assignment. One of the websites was the fake one she had created. And though it's hard to believe, some of the kids copied the fake facts directly from her website and handed them in on their papers.

Now I'm not saying that your teacher will try to trick you like that, but it's a good example of why you shouldn't believe everything you see on the Web. (But you *can* believe everything you read in a book. Okay, I gotta go now—my spaceship is waiting to take me to Saturn.)

IT'S .COM-PLICATED

Did you know that different countries have different website addresses? For example, to access a website in America, you often end with ".com." But in Brazil, it's ".br," Australian websites frequently use ".au," and in Denmark, it's ".dk." And there are more than 200 others.

How do I know all this? I'm ".smart."

MEET DR. SHIRLEY ANN JACKSON

Dr. Shirley Ann Jackson was not just an inventor; she was a trailblazer. Dr. Jackson was one of the first African American students to attend MIT (the Massachusetts Institute of Technology), and there was only one other woman in her undergraduate class. What's more, in 1973, she became the first African American woman to receive a PhD (Doctor of Philosophy) degree at the university. A trailblazer indeed.

In 1976, Dr. Jackson joined AT&T Bell Laboratories. Using the knowledge of particle physics she'd gained at MIT, she researched ways to improve telecommunications—the transmission of voice, writing, images, and video over long distances. Remember, this was many years before people had the Internet, texting, FaceTime, and other ways they now use to keep in touch.

Dr. Jackson conducted studies in optical physics, theoretical physics, and solid-state and quantum physics. What are those? I have no idea—*I* don't have a PhD from MIT. But partially because of Dr. Jackson's successes, you can now look it up for yourself online.

The inventions to which Dr. Jackson greatly contributed include fiber optic cables, which provide faster and more reliable transmission of voice, images, and data. She was also instrumental in the introduction of—and improvements to—the touch-tone telephone, the fax machine (which was *very* popular for sending documents in the 1980s and 1990s, though not so much today), call waiting (so that when you hear a *beep* during a phone call, you know someone else is trying to reach you), and caller ID (the feature that lets your telephone alert you to who's calling).

Dr. Jackson's research also gave us the solar cell, a device that converts light from the sun into useable electricity. More and more, you can see these on top of houses; the owners have installed them to harness the power of the sun to provide electricity to their homes. It's said to be better for the environment (and saves money too).

Dr. Jackson has given the world much to appreciate. And in 2016, the world showed its appreciation as President Barack Obama awarded her the National Medal of Science, America's highest honors for contributions in science and engineering. Today, Dr. Jackson is the president of Rensselaer Polytechnic Institute in upstate New York.

SOME DEFINITE DEFINITIONS

Okay, I felt guilty not defining optical physics, theoretical physics, and solid-state and quantum physics for you. So, I looked them up, and here are some definitions.

Optical physics is the study of the fundamental properties and behavior of light.

Theoretical physics is the use of mathematics to describe aspects of nature.

Solid-state physics is the study of the physical properties solids.

Quantum physics is . . . (I'm sorry, I looked it up and I still don't understand it. My apologies!)

THEY CALL ME 'FRANCE-Y PANTS!'

I speak French pretty well. And thanks to fiber optic cables, I now fully understand what my friends in Paris are saying on the phone. But before those cables were in use, it was almost impossible to figure out their statements.

For example, “Comment allez vous?” translates into English as “How are you?” But I thought they were saying, “Comment ballet shoes?” which means “How ballet shoes?” (Which means nothing.)

Or they'd say, “c'est nuageux,” which means, “it's cloudy.” But I thought they were saying, “say nauseous.” So, I would say, “nauseous.” Over and over, and over. And they'd hang up.

You can understand why I'm a big fan of fiber optic cables.

FACT VS. FICTION

Fact: It can take up to nine years to go through college and graduate school for a PhD in the field of physics.
Fiction: When you're in graduate school, you can do your homework on an Etch A Sketch.

Fact: Although the recipient of a PhD earns the designation of "Dr.," he or she cannot fix a broken arm.
Fiction: Interestingly, the recipient of a PhD *can* often fix a delicious tuna sandwich.

Fact: Tuna sandwiches go very well with milk.
Fiction: We get milk from frogs.

IT'S MEDAL TIME!

Dr. Jackson received a well-deserved medal for her work in science. As someone who's reading this book, I think *you* deserve a medal as well. Believe me—it's meant just for you, and it's only in *this* copy of the book. In other copies, this page is blank. Totally blank. You do believe me, don't you? Well, don't you?

And listen, the only reason it doesn't have your name on it is because I don't actually know your name. So, I'll call you "Hey, Kid." Okay, thanks.

HEY, KID,
A GREAT
FRIEND
AND A GREAT
READER

MICROWAVE
POPCORN
TASTY

MEET PERCY SPENCER

You get home from school and you're hungry. Starved. It feels like a popcorny kind of day, and since you don't live in a movie theater, you toss some pre-packed kernels in a boxlike machine. Two minutes later . . . you've got hot, buttery, delicious popcorn!

That quick, yummy moment was brought to you by Percy Spencer, the inventor of the microwave oven.

As a boy, Mr. Spencer was always trying to figure out how things worked. Upon joining the U.S. Navy at age eighteen, he began learning all he could about radio and wireless technology as well as complex math and science.

After completing his military service in the mid-1940s,

this self-educated man took a job with the Raytheon Company, heading up their power tube division. He certainly didn't set out to create the microwave oven, but as with many great inventors, genius struck when least expected.

See, Mr. Spencer was in the Raytheon plant, working in a lab, testing magnetrons—high-powered vacuum tubes inside radars—which produced microwaves. When he stood near a magnetron, the peanut-cluster candy bar in his pocket got warm and melted. That triggered something inside Mr. Spencer's mind; he recognized that food cooks quickly when exposed to low-density microwave energy. He decided to try this "experiment" again, this time with popcorn. As you can guess, the kernels popped! Then he cooked an egg (okay, so it exploded, but it *did* cook).

After conducting more research, Mr. Spencer invented, refined, and received a patent for what we now know as the microwave oven. However, the 1947 version of the oven was as large as a refrigerator; called the Radarange, it stood nearly six feet tall and weighed 750 pounds. The Radarange (a name combining the words *Radar* and *Range*) cost about $2,500 (roughly twenty-five times more than today's microwave ovens do) and was purchased mostly for use in restaurants.

About twenty years later, a smaller, more affordable (about $500) microwave oven was offered to consumers for home use.

According to the U.S. Census Bureau, more than 98 percent of American households now own a microwave oven. And Mr. Spencer's other work with microwave technology impacts many lives on a daily basis. Since they're able to pass through rain, snow, and clouds, microwaves are used with satellites that track weather conditions. Microwaves are also the basis for radar guns that allow police to determine if a car is speeding.

By the way, you're probably thinking that Mr. Spencer became a multi-millionaire thanks to his invention of the microwave oven. But nothing could be further from the truth; because he was an employee of Raytheon, he was given a one-time payment of . . .

. . . two dollars.

That's right; all the inventor of the microwave oven received was just enough to buy a box of microwave popcorn. But he made a big difference in the world, and that was surely personally rewarding.

AS A MATTER OF NON-FACT . . .

According to what I've read online, there are 79,312 kernels of popcorn in the average bag of microwave popcorn.

NO, THERE AREN'T—*I told you not to believe everything found online.*

Actually, there are approximately 450 kernels in a typical bag of microwave popcorn. And that's something you *can* believe. Or, you can count them yourself . . . but wait until they cool down if you do.

CONSIDER THE SAUCE

In an effort to duplicate Mr. Spencer's amazing breakthrough, I put a slice of leftover meat loaf in my pocket and walked back and forth for twenty minutes in the microwave oven section of my local appliance store.

The meat loaf didn't cook; rather, the meat loaf sauce ran down my leg and onto the carpet of the store. The store manager asked me to leave.

My discovery? Don't ever walk around with meatloaf in your pants pocket.

EVEN MORE IMPORTANT MATH! EVEN MORE IMPORTANT MATH! EVEN MORE IMPORTANT MATH!

Generally, microwave ovens cook things four times faster than traditional wall ovens. So, if a person cooked every meal in the microwave oven, what would take an hour in a regular oven would take only 15 minutes in that microwave. Using that same math, what takes 96 hours in a regular oven would only take 24 hours in a microwave. Therefore, the person using the microwave would have 3 extra days in their life compared to the regular oven user. Multiply that by all the times people cook, and if the oven user lives until 90 years old, the microwave user would live to 360.

This, of course, is total nonsense. It seemed right to me until I typed it, and then I realized it was just plain silly. And if you don't like your silly just plain, then add a little salt and pepper, perhaps some mustard or hot sauce, then have someone put it in the microwave for you. It's bound to be delicious . . . and it's sure to cook quickly!

MEET PATSY SHERMAN

Your Mom and Dad have reminded you 4,234,825 times: no drinks on the couch.

But they're busy upstairs, and just this one time wouldn't hurt. Would it?

It would! No matter how careful you've vowed to be, you somehow end up spilling apple juice all over the cushions!

It could be a mess. It could be a calamity. But . . . it's a miracle! The apple juice doesn't sink into the fabric and doesn't instantly stain the cushions.

Why? One word: *Scotchgard™*!

More specifically, Scotchgard™ Textile Protector, a

stain-repelling product pioneered by Patsy Sherman in the early 1950s.

Ms. Sherman, then a research chemist at the 3M Company, was working with another chemist to develop rubber for jet fuel hoses. (His name was Samuel Smith, and he held thirty patents of his own during the four decades he spent at 3M.)

When Ms. Sherman and Mr. Smith accidentally spilled a chemical compound on their assistant's white tennis shoe, they noticed that it didn't stain the shoe. And, incredibly, no matter how hard they tried, they couldn't remove that compound; it repelled water and simply wouldn't wash off. It even made the canvas shoe stain resistant.

Ms. Sherman and Mr. Smith knew they had made quite a discovery—one that other chemists had long considered scientifically impossible. They worked for a few years to perfect their innovation, which in 1956 was sold to consumers as Scotchgard Fabric Protector. Before long, there were Scotchgard products designed specifically to treat wool, upholstery, carpeting, and other fabrics.

By the way, you might be wondering why they named it Scotchgard. I wondered that, too, until I realized that Scotch™ Tape is also a product made by the 3M Company; I suppose they wanted people to know it was another reliable 3M product.

Today, you can buy Scotchgard cans to spray onto your furniture, car seats, and other belongings. You can also buy sofas, chairs, and many household items that are pre-treated with the protectant.

Incredibly, while Ms. Sherman's breakthrough formula was being tested in 3M's textile mill, she was not allowed inside. Back in the 1950s, Ms. Sherman was one of the only female chemists in America; at that time, women were forced to remain outside of the mill and wait for performance results.

For her work on Scotchgard as well as on the many other patents she holds, Ms. Sherman was inducted into the National Inventors Hall of Fame. "Anyone can become an inventor as long as they keep an open and inquiring mind and never overlook the possible significance of an accident or apparent failure," said Ms. Sherman.

What a wild, life-changing story: the accidental spill *Ms. Sherman* had in the lab made it possible for the rest of us to stay out of trouble when *we* spill something on the furniture. We salute you, Ms. Sherman; your record as a chemist is unstained, and so are our couches.

MURRAY DID IT!

You now know that the 3M Company gave the world Scotch™ Tape and Scotchgard™ products. But did you know that "3M" stands for Minnesota Mining and Manufacturing Company? That's true; they've been around for more than a century and are headquartered in Minnesota.

Did you also know that there was another company, called 1M, that was just a guy named Murray, working in his basement (also in Minnesota)? He accidentally invented a couch that comes *pre-stained*—it was designed for kids who wanted a built-in excuse whenever they made a mess: "Don't blame me—Murray did it!"

It was also intended for people who were really very tidy but wanted their visitors to think they're stain-making slobs.

Murray's couches were available in mud stains, food stains, paint stains, and unidentifiably disgusting stains.

He never sold a single pre-stained couch.

Because I made up the whole story of Murray, his company, and his product line.

So put your money away—as terrific as those couches may sound, they're simply not available for purchase.

1M
1M

AN ODE TO PATSY SHERMAN AND SCOTCHGARD

The chocolate ice cream pop didn't leave a stain.
Neither did the tapioca, the grape juice, or the oranges we got from Cousin Rachel in Tampa.
In fact, the only thing no one can seem to remove from our plush comfy chair in the den, is . . .
. . . the one and only man we call Grampa!

NO CAN DO

I had a terrible nightmare. I got a stain on my can of Scotchgard, and to make sure it never happened again, I sprayed Scotchgard on that can of Scotchgard. Then, to make sure that the second can of Scotchgard didn't get stained, I got a third can and sprayed it on the second. This continued on and on, until I had a room full of Scotchgard cans!

I told my son Nathan about the nightmare. He told me it was ridiculous; I could've sprayed the first can with the second, and the second with the first. No need for cans 3 through 5,436.

"Besides," he told me. "Who care if a can of Scotchgard gets a stain on it

He has a point. I'm going back to sleep.

A FINAL WORD FROM ALAN KATZ

Well, that's what I have to say about some of the Awesome Achievers in technology. I hope that you've learned a lot. I hope that you've had a ton of fun. And I hope that you'll take the time to find out even more about the incredible people I've discussed in this book.

Most of all, I hope you'll keep reading. Because I'd bet if you asked any of the Awesome Achievers in this book what they considered the secret to their success, they'd each say it was knowledge. And you can get knowledge from books.

Make sense? I sure hope so. And, hey, if you have ideas for future Awesome Achievers books, please write to me at awesomeachieversbooks@gmail.com.

Thanks. And before you go . . . please clean up the meatloaf sauce that's dripping out of your pants pocket!

Bye!

NOTES